Raising Beef Cattle
Agricultural Bulletin 138

by Pennsylvania School of Agriculture

with an introduction by Jackson Chambers

This work contains material that was originally published in 1916.

This publication is within the Public Domain.

This edition is reprinted for educational purposes
and in accordance with all applicable Federal Laws.

Introduction Copyright 2018 by Jackson Chambers

Self Reliance Books

Get more historic titles on animal and stock breeding, gardening and old fashioned skills by visiting us at:

http://selfreliancebooks.blogspot.com/

Introduction

I am pleased to present another title in the "Cattle" series.

The work is in the Public Domain and is re-printed here in accordance with Federal Laws.

As with all reprinted books of this age that are intended to perfectly reproduce the original edition, considerable pains and effort had to be undertaken to correct fading and sometimes outright damage to existing proofs of this title. At times, this task is quite monumental, requiring an almost total "rebuilding" of some pages from digital proofs of multiple copies. Despite this, imperfections still sometimes exist in the final proof and may detract from the visual appearance of the text.

I hope you enjoy reading this book as much as I enjoyed making it available to readers again.

Jackson Chambers

INVESTIGATION ON RAISING BEEF CATTLE

BY B. O. SEVERSON[*]

INTRODUCTION.

This investigation was begun Dec. 1, 1911, by W. A. Cochel, who was then in charge of the Animal Husbandry Department of The Pennsylvania State College and later succeeded by W. H. Tomhave. The results of the first winter's work were reported by W. A. Cochel in Bulletin 118 of this Station. Two detailed reports on the investigation have been made by W. H. Tomhave and B. O. Severson, the first in the Annual Report of The Pennsylvania State College in 1912-13, and the second in 1914-15.[1] The average results of the first three years of the experiment are reported in the following pages. The investigation, however, is still in progress.

Pennsylvania[2] has a total acreage of 28,692,480 of which 16,018,961 acres are not considered improved farm lands. Of this unimproved land 5,893,313 acres are in farms. The total amount of improved farm land in the State is 12,673,519 acres. How shall this vast unimproved area be used? The number of principal grazing animals in the State in 1910 included 1,586,519 cattle and 883,000 sheep. The farmers of Pennsylvania annually feed over 200,000[3] steers. Would it be possible to raise cattle on the rougher and less fertile lands of the State and thereby supply the demand for "feeders" in the sections where land is higher in value and better adapted to intensified farming? These and other related questions in farm management, such as soil fertility, labor and farm equipment, were responsible for this investigation.

---

[*] The details of this investigation were reported in a thesis for the degree of Master of Science in June, 1915, by the author.

[1] Reprints showing details of the investigation including costs will be mailed on request. First reprint—Annual Report, Pennsylvania State College 1912-13, now available. Second reprint—Annual Report, Pennsylvania State College 1914-15, available Jan. 1, 1917.

[2] United States Census Report, 1910.

[3] Pennsylvania Department of Agriculture, Harrisburg, Pa., Bulletin 235—Lancaster market disposes of 120,000 "feeders" annually.

OBJECTS OF THE INVESTIGATION.

The objects were to determine the possibilities of raising beef cattle in Pennsylvania; whether the demand for "feeders" could be met profitably; the cost of maintaining breeding stock; the cost of raising growing and breeding beef cattle; the cost of finishing beef cattle for market; the value of silage as a sole roughage, and of cottonseed meal as a supplementary feed; and finally, to study details of management in feeding and breeding of beef cattle for profitable production.

Fig. 1.—A Badly Eroded Field on One of the College Farms.

PLAN OF THE INVESTIGATION.

Twenty pure-bred beef cows were used in this experiment. Ten Shorthorns composed Lot I, and ten Aberdeen-Angus composed Lot II. The object was not a breed comparison, but to have cattle which would represent standard breeds of beef type. During the summer months the breeding cows and growing stock were on pasture with no extra feed in the form of grain or roughage. During the winter months corn silage was the sole roughage fed

Fig. 2.—Same Field as Shown in Fig 1, After Being Sown for Hay and Permanent Pasture.

to the breeding and growing stock. Cottonseed meal was fed, in addition, at the rate of 1 pound per cow, daily, and at the rate of three pounds per thousand pounds live weight, daily, to the growing stock. The fattening stock was fed in accordance with methods proven by experiment to be profitable at this Station.[4] The calves ran with their dams and were weaned at seven to eleven months of age. The cows were bred to pure-bred beef bulls, such cows as failed to breed being removed from the experiment and others substituted from the college herd.

DESCRIPTION OF BREEDING CATTLE.

The ten pure-bred Shorthorn cows[5] used in this investigation were purchased in Mercer County, Pennsylvania, in the fall of 1911. They were in relatively thin condition of flesh; none had calves at their sides, nor were they bred when purchased. This fact must be considered when reading this report. Some of these cows failed to become pregnant, while in other cases the calves were dropped after the investigation had been in progress for a year or more. These cows were of a "useful kind" and not of show-yard excellence.

---

[4] Pennsylvania State College Bulletins 118, 124 and 133.

[5] Annual Report, Pennsylvania State College, 1912-13, page 108.

Fig. 3.—Lot I, Shorthorn Cows and Calves Just Out of Winter Quarters.

The ten Aberdeen-Angus cows were taken from the college herd.[6] All of them were tried breeders and only one was considered doubtful. They were in good condition of flesh when placed on the experiment. With respect to relative individual merit they were superior to the cows in the Shorthorn lot, but would be described by the pure-bred breeder as a "useful kind." The object was to have cattle of insured beef breeding for the production of beef calves that would make certain the early maturity, form, quality, and finish recognized as essentials in an animal destined for the butcher's block.

The bulls[7] used were registered animals of the Shorthorn and Aberdeen-Angus breeds. Their individual merit was higher than that averaged by the breeding cows of either lot.

FEEDS FED TO BREEDING COWS.

"This experiment was undertaken to find some feed that would produce a larger amount of food nutrients per acre than hay, and be as well adapted to the feeding of breeding cattle when supple-

---

[6] Annual Report, P. S. C., 1912-13, Page 108.
[7] Described in Annual Report, P. S. C., 1912-13 and 1914-15.

mented with a minimum amount of concentrated feeds which would furnish the digestible nutrients not provided by the farm-grown crop. Corn silage was selected as most nearly meeting these conditions. It is adapted to a wider range of soil conditions than any other crop except grass; it produces a larger amount of food nutrients to the acre, is palatable, succulent, easily grown, harvested with comparative ease and can be stored at less expense for buildings than any other crop. In addition to these advantages, there is no other form in which the corn crop will be entirely consumed by live stock; thus, it increases in value by being placed in the silo. Cottonseed meal was used as a supplement because of the fact that protein, in which corn is quite deficient, could be secured in this form cheaper than from any other source, and because of the large percentage of protein that could be fed in very small quantities of the meal, thus reducing the expense of transportation and labor in feeding. Previous investigation had also shown that the laxative tendency caused by heavy feeding of succulent feeds is materially reduced by the addition of cottonseed meal to the ration."[8]

WINTER QUARTERS.

All the cows were sheltered during the winter months in protected but open quarters. Experimental results at this and other stations had shown that open sheds[9] made desirable quarters for fattening steers. The sheds used consisted of a building boarded up on three sides with an open side to the south. Adjoining the shed was an open yard 25 feet deep, and thus was formed an enclosure 40 feet deep and 30 feet wide. Each lot provided shelter for ten cows and their calves. During the third winter the basement of a barn with the south side open was used; this provided the same type of shelter as that used during the first two winters. The calves were placed on a growing ration during the second winter, and the calves fattened for beef the third winter were kept in the basement of a barn because no open quarters were available. At all other times the growing stock was sheltered in quarters similar to those used by the cows. Special care was taken to keep the quarters well bedded and dry. Water was kept before the cattle in galvanized iron troughs.

---

[8] Pennsylvania Experiment Station Bulletin No. 118.
[9] Pennsylvania Experiment Station Bulletin No. 102.

WEIGHTS AND RECORDS.

Individual weights of all the cattle were taken on three consecutive days, on full feed and water at the beginning and close of each winter and summer period. During the winter weights were taken every four weeks on three consecutive days,—on the first and third, group weights and on the second, individual weights. Also in the middle of each four-week period group weights were taken. During the summer individual weights were obtained every four weeks. Calves were weighed at birth and at the regular weighing periods thereafter. Records were kept of all feeds fed and refused during the winter. The time of service and date of calving were recorded. Records were kept on the behavior and condition of individuals in the herd.

WINTER FEEDING AND MANAGEMENT OF BREEDING COWS.

The ration fed during the three winters was the same for the Shorthorn and the Aberdeen-Angus lots. Corn silage was fed twice daily in quantities such as would be cleaned up in one hour after feeding. Cottonseed meal was fed twice daily at the rate of one-half pound to each cow, per feed. The cottonseed meal was distributed over the silage in the manger. Salt in limited amounts was placed in the manger three times each week.

All the cows in each lot were fed together in a manger thirty feet long and two feet wide. During cold weather when cows were due to calve, as shown by the normal period of gestation, 283 days, or by outward symptoms, they were separated from the herd and placed in quarters that would give them more warmth, comfort and solitude. They were placed in these quarters a day or two before calving but were allowed the same feeds and as much exercise as possible. From birth the calf was gradually accustomed to exposure so that at the end of six days it might be returned with its dam to the herd. Special attention was given to the cow, with reference to the removal of the placenta (cleaning) within twenty-four hours after parturition, and to the udder to prevent caking or other injury. Cows after parturition were usually milked every other day for a week or more, the frequency and extent of milking being the greater when the milk produced was more abundant and the calf unable safely to utilize the supply. In case of the calves, precaution was taken to avoid scours, caused by consumption of too much milk. This was overcome by "stripping" the cow night and morning, a measure which prevented the calf from getting the milk richest in butter-fat. The calves were given no feed nor care other than the nourishment received from their dams. The cows were bred as soon as they came in heat.

SUMMER MANAGEMENT.

Fig. 4.—Lot II, Aberdeen-Angus Cows and Calves on Summer Pasture.

The breeding cows and their calves were placed on pasture, during the three summers, on April 20, 1912, April 25, 1913 and April 26, 1914. At State College, if pastures are not grazed too short in the fall, they will be satisfactory for grazing by April 25. Salt was available at all times. Special attention was given to get all cows bred. It became noticeable that the cows showed a tendency to come in heat most regularly in May and June. No shelter was provided on pasture except during the cold rains of spring and fall. At such times young calves were sheltered at night. The pasture area contained about 118 acres. Other stock consisting of sheep and other beef cattle were kept on the same pasture. Overstocking was guarded against to insure the building up of a good permanent sod. The pasture during the summer of 1913-14 was greatly improved over the stand of 1912. Fewer weeds were found, owing to the efficiency of the sheep in eradicating them; the sod was denser, and the clover more abundant. The area was hilly and better adapted to a permanent pasture than to cultivated crops, because of the tendency to severe soil erosion after cultivation.

RESULTS OF THREE YEARS[10]

Dec. 1, 1911, to Nov. 25, 1914.

The cost of maintaining a beef breeding herd and of raising and fattening young stock is reported herewith. Furthermore this report shows the behavior of the breeding animals in feeding and breeding, and the development of their calves in growing and fattening periods.

Table I.—Summary of Maintenance of Beef-Breeding Cows.
(Average of Three Winters.)

	Lot I 10 Shorthorns	Lot II 10 Amberdeen-Angus
Average length of winter	154.6 days	154.6 days
Total initial weight per cow	1179.546 lbs.	1142.899 lbs.
Total final weight per cow	1267.531 "	1198.186 "
Total gain weight per cow	87.986 "	55.287 "
[1]Total feed per lot		
Corn silage	91010.41 "	89201.91 "
Cottonseed meal	1546.66 "	1546.66 "
[1]Average daily feed per cow		
Corn silage	58.83 "	57.63 "
Cottonseed meal	1.00 "	1.00 "
[2]Average air-dry matter, per cow daily in feeds consumed	21.88 "	20.95 "
Cost of feeds		
Corn silage @ $3.50 per ton	$159.27	$156.19
Cottonseed meal @ $30.00 per ton	23.22	23.22
[2]Average cost of feed per cow	18.25	18.14
[3]Bedding used per cow, 1090 lbs. @ $8.00 per ton	4.35	4.35
[4]Labor of feeding	2.33	2.33
[3]Value of manure per cow, 9758 lbs. @ $1.50 per ton	7.32	7.32
Net cost of wintering a cow	17.61	17.50

1. Does not include 810 lbs. alfalfa and 75.5 lbs. of grain fed to cow and calf in Lot II during third winter.
2. Include feeds in (1).
3. Based on weights taken first winter, reported in Pennsylvania Experiment Station Bulletin 118.
4. Labor cost 15 cents per hour.

Table I shows the average results obtained in the maintenance of beef breeding cows during the winters of 1911-12, 1912-13 and 1913-14. The average length of the winter period was 154.6 days, or about five months. Each winter each lot showed gains at the

---

[10] Annual Reports, The Pennsylvania State College, 1911-12, 1913-14, and 1914-15.

conclusion of the feeding period. The relatively greater gain in Lot I was due to the thinner condition of flesh which the cows showed at the start of the experiment. Slightly more feed was consumed by the Shorthorns than by the Aberdeen-Angus cows, as indicated by the air-dry matter in feeds consumed. In Lot I, 21.88 pounds and in Lot II, 20.95 pounds of air-dry matter in feeds fed were consumed per cow, daily. Each cow required 1,090 pounds of bedding during the winter period. The amount of manure produced was 9,785 pounds per cow, which more than paid for the bedding and labor involved per cow even when manure was figured at the low valuation of $1.50 per ton. The cost of wintering per cow was $17.61 in Lot I and $17.50 in Lot II.

Table II.—Summary of Pasturing Beef-Breeding Cows During Summers of 1912, 1913 and 1914.

	Lot I Shorthorns	Lot II Aberdeen-Angus
Average length of summer period	210.3 days	210.3 days
Total initial weight per cow	1208.375 lbs.	1174.895 lbs.
Total final weight per cow	1258.521 "	1189.519 "
Total gain in live weight per cow	50.146 "	14.725 "
[1] Interest on land area on which cows grazed	$56.00	$56.00
[2] Interest on land area on which calves grazed	6.31	13.16
[3] Labor cost	8.03	8.03
Net cost per cow	7.034	7.710

1. Two acres allowed for a mature cow during pasturing season. Owing to the rough and broken land which made the pasture area suitable only for grass, interest at 5 per cent. was allowed on the value of the pasture area. The farm was valued at $56 per acre, a fair return for land otherwise unproductive.
2. Calves after four months of age, while nursing with dams, were allowed one acre of land for grazing during the season.
3. Labor cost 15 cents per hour.

Table II shows the results of pasturing beef-breeding cows. The average length of the summer period was 210.3 days or seven months. Cows in both lots showed gains, at the conclusion of the summer period, of 50.146 pounds per cow in Lot I, and 14.725 pounds per cow in Lot II. The greater gains in Lot I were due to the introduction of several growing heifers to replace non-breeders, and also to the fact that the Angus cows nursed a greater number of calves. The total cost of pasturing was $7.034 per cow in Lot I, and $7.71 in Lot II. The difference between the lots is due to the greater number of calves raised by cows in Lot II. The cows on pasture received no feeds aside from grass.

Table III.—Yearly Cost of Maintaining Beef-Breeding Cows.
Average of Three Years, Dec. 1, 1911, to Nov. 25, 1914.

	Lot I Shorthorns	Lot II Aberdeen-Angus
Cost of wintering per cow	$17.61	$17.50
Cost of summering per cow	7.03	7.71
[1] Service of Sire	2.00	2.00
Interest on value of cow	5.40	5.40
Interest on equipment for shelter and feed	1.50	1.50
	$33.54	$34.11

1. Based on a sire worth $200, for a herd of 50 cows annually, allowing $100 for feed, labor and interest on value and equipment.

Table III shows the yearly cost of maintaining beef-breeding cows for beef production. The cost of wintering for 154.6 days was more than twice as much as that of pasturing the cows for a period of 210.3 days. This fact emphasizes the importance of paying heed to pasture in order that it shall maintain the cattle for the longest possible time. The cows were valued at $90 per head. Interest at six per cent. was reckoned on the value of each cow. The cows that were sold as non-breeders brought from $6 to $7 per cwt. on local and Pittsburgh markets. In each case this value of the cows for beef would be greater than the cost of raising them to a breeding age in the herd. The interest on shelter, silo and equipment for feeding was based on actual valuation at five per cent. interest. The total annual cost of maintenance for a breeding beef cow in Lot I was $33.54, and in Lot II $34.11. No distinction can be drawn between the cost of maintenance of breeding cows of like merit in the Shorthorn and Aberdeen-Angus breeds under the conditions of this investigation.

RESULTS OF BREEDING[11]
Dec. 1, 1911, to Nov. 25, 1914.

The results presented here do not show the relative merits of the Shorthorn and Aberdeen-Angus breeds, but indicate rather the necessity of care in the selection of breeding animals for the establishment of a beef herd. None of the Shorthorn cows were pregnant when the experiment began Dec. 1, 1911, and nothing authentic

---
[11] Tables 20 and 21, Annual Report of The Pennsylvania State College 1913-14, show detailed results.

Fig. 5.—Oneida Chief 364,944, the Shorthorn Bull.

was known relative to their breeding merit. In the case of the Aberdeen-Angus cows, all of them had previously dropped calves in the college herd, and a majority of them were pregnant when the experiment began. All the bulls used were pure-bred. In two instances cross-bred calves were raised.

Table IV.—Summary of Breeding Results.
Dec. 1, 1911, to Nov. 25, 1914.

	Lot I Shorthorns	Lot II Aberdeen Angus
Percent. normal calves born	46.66	70.00
Percent. male calves	76.	36.36
Percent. female calves	24.	63.64
Number of deaths (normal calves)	1	1
Number of calves aborted	4	2
Average period of gestation	285 days	284.5 days
Shortest period of gestation	249 "	259 "
Longest period of gestation	313 "	314 "
Number of services per calf	1.63 times	1.43 times
Average weight normal calves at birth	75.27 lbs.	68.36 lbs.
Highest weight normal calves at birth	100. "	100. "
Lowest weight normal calves at birth	42. "	40. "

This table is significant because it emphasizes one of the most essential factors in the successful management of a beef herd, namely, the selection of cows that are regular breeders. In the Shorthorn lot, only 46.66 per cent. normal calves were born, while in the Aberdeen-Angus lot 70 per cent. normal calves were born. In Lot I, in seventeen cases where sex was determined, 76 per cent. were males and 24 per cent. were females. In Lot II, out of 22 calves, 63.64 per cent. were females and 36.36 were males. The loss of one normal calf in each lot was due to a case of exposure in Lot I, and the death of a twin in Lot II that lived but a few hours. Four abortions occurred in Lot I, two of which happened during the summer period and two during the winter period. In no case were there symptoms of contagious abortion in Lot I. In two cases the winter ration had no connection with the abortion since one abortion occurred in July and the other in September. One cow aborted five days after the beginning of the second winter period and another in mid-winter. One abortion occurred after the time covered by this investigation, as herein reported. The Pennsylvania Live-Stock Sanitary Board, at its laboratories, made a

Fig. 6.—Jower 119,029, the Aberdeen-Angus Bull.

complement fixation test of this case with negative results. In Lot II, the two abortions occurred during the first winter. These two cases were suspicious but no positive evidence as cases of contagious abortion was obtained. The average gestation period was 284.5 days in Lot I and 285 days in Lot II. The average number of services for each cow that became pregnant was 1.63 times in Lot I and 1.43 times in Lot II. In Lot I, seven of the original ten cows were discarded as non-breeders. In Lot II one was discarded for failure to become pregnant and two for failing to breed after each had dropped one calf. The Shorthorn calves averaged 75.27 pounds, and the Angus calves averaged 68.36 pounds, at birth.

RESULTS OF GROWING STOCK.

After the calves were weaned they were separated from the breeding cows. During the winter they were fed corn silage to the limit of appetite and cottonseed meal at the rate of three pounds per thousand pounds live weight, daily.

Table V.—Summary of Results From Growing Calves After Weaning.[12]

Average length of nursing period	10½ months
Average weight at weaning	565. lbs.
Average weight Shorthorn calves at 12 months	671.6 "
Average weight Aberdeen-Angus calves at 12 months	588.8 "
Average weight Shorthorn heifers at 24 months	991.6 "
Average weight Aberdeen-Angus heifers at 24 months	965.8 "
Average feed consumption per calf during winter	
Cottonseed meal	363.26 "
Corn silage	5793.33 "
[1]Cost of winter feed for calf having weight averaging 573.5 lbs.	$15.67[3]
[2]Cost of summer pasture per growing animal having average weight of 831.2 lbs.	4.20[3]
Average daily gain during winter period per calf	1.243 lbs.
Average daily gain during summer period per head	.693 "

1. Cottonseed meal valued at $30 per ton, corn silage $3.50 per ton.
2. One and one-half acres of pasture was allowed for each growing animal.
3. Labor cost was charged on breeding stock to cover the cost of growing stock.

Table V shows the summary of results in feeding growing beef calves during the last two years of the experiment. Calves at weaning time were placed in this lot of growing stock which

---

[12] Annual Report, The Pennsylvania State College, 1914-15.

Fig. 7.—Growing Calves as They Appeared When Removed From Winter Quarters.

included a total of 21 for the two winters and 18 for the two summers. The average weights of heifers at 24 months of age were based on those of three Shorthorns and six Aberdeen-Angus heifers. The labor involved was not charged, since the charge made against the breeding stock covered the cost involved by the growing stock.

Table VI.—Value of Calves at Twelve Months at Varying Percentages.

	Cost of Calf[2]	Value of Calf[1]
Percent calves raised in herd		
100 percent	$38.26	$50.40
90 percent	42.08	50.40
80 percent	45.91	50.40
70 percent	49.73	50.40
60 percent	53.56	50.40
50 percent	57.38	50.40

1. Valuation based on $8 per cwt.
2. Cost was based on cost of maintenance per cow and the feed consumed by calf after weaning.

Calves at weaning time cost as much as it costs to maintain a beef breeding cow for one year, provided 100 per cent. calves are raised. In a well-selected herd under good management 80 per cent. calves, annually, should be considered the lowest level. In this investigation at 70 per cent., the cost of a calf at 12 months was practically the same as its value. When 80 per cent. calves are raised they cost $40.60 each at weaning, at an average weight of 565 pounds.

The cost of raising heifers to a breeding age, that is 30 months, was $76. They weighed more than 1,050 pounds at this age and their market value at $8 per cwt. would make them worth $84. A heifer in good condition for breeding can be raised for less than her market value as beef when 80 per cent. calves are raised.

FATTENING STEERS.

Five steers were used for fattering purposes during the winter of 1913-14. As shown by experimental results[13] corn silage as a sole roughage, when supplemented with cottonseed meal and corn, proved to be a cheap and efficient feed for fattening steers. However, for fattening cattle designed as "yearlings"—that is, steers less than 24 months of age weighing about 1,000 pounds—into marketable condition, experience has shown that some legume hay, as clover or alfalfa hay, brings good results, in combination with silage and grain.

The ration fed consisted of corn silage in amounts to satisfy the appetite and cottonseed meal at the rate of three pounds per thousand pounds live weight, daily, with corn-meal in limited amounts at first, but gradually increased to a full feed after two months of feeding. Corn-meal was fed because the quarters would not permit pigs to follow the steers. Hogs should follow cattle wherever possible, since one to two pounds of pork are produced for every bushel of shelled or ear corn consumed by steers.[14]

---

[13] Pennsylvania Experiment Station Bulletins 102, 118, 124, and 133, and Annual Report, The Pennsylvania State College, 1914-15, on "Methods of Steer Feeding."

[14] Bulletin 146, Purdue University.

Fig. 8.—The Fattening Steers as They Appeared When Sold.

Table VII.—Summary of Results in Fattening Five Steers.
Nov. 25, 1913, to April 20, 1914. 157 Days.

Initial weight	3609.99 lbs.
Final weight	5145.00 "
Total gain	1535.01 "
Average daily gain per head	1.95 "
[1]Initial cost of 5 steers	$220.44
[2]Final value of 5 steers	409.33
Total amount of feed fed	
Corn-meal	7370.50 lbs.
Cottonseed meal	1992.00 "
Corn silage	16490.00 "
[3]Total cost of feeds	$150.95
Average air-dry matter in feeds consumed, per head, daily	17.89 lbs.
Total profit in feeding	$38.94
Average profit per head	7.79

1. Based on results of experiment in which 80 per cent. calves are raised.
2. Sold for $8.20 per cwt., with 3 per cent. shrinkage allowance in shipment.
3. Values of feed; cottonseed meal $30 per ton, corn-meal 70 cents per bushel, and corn silage $3.50 per ton.

The valuation of the steers was based on the prospect of 80 per cent. calves being raised. These steers cost $44.18 each, or

$6.10 per cwt.[15] In comparison, the college purchased 60 steers in West Virginia at $7 per cwt. in the fall of 1913. After paying commissions, freight, allowing for shrinkage, and preliminary feeding before being placed in the feed lot, they cost $7.45 per cwt. The steers raised at a cost of $6.10 per cwt. showed better quality, breeding and condition than those that cost $7.45 per cwt. A profit of $7.79 per head was made on the steers fed in this experiment notwithstanding the fact that the spring of 1914 proved disastrous for steer feeding profits throughout the country.[16]

CONCLUSIONS.

1. Beef-breeding cows were maintained in good condition of flesh on a ration composed of corn silage in amounts to satisfy appetite, supplemented with one pound of cottonseed meal, daily.

2. Weaned calves made satisfactory gains on corn silage supplemented with cottonseed meal at the rate of three pounds, daily, per thousand pounds live weight.

3. The normal calves in both lots were vigorous at birth.

4. There was no distinction between the Shorthorn breed and the Aberdeen-Angus breed in the utilization of feeds, cost of maintenance and breeding qualities. (The low percentage of calves raised in the Shorthorn lot was due to the large proportion of non-breeders in this lot at the beginning of the experiment).

5. About four and one-half tons of corn silage was consumed by each cow during the winter; the exact average was 9,101 pounds in Lot I and 8,920 pounds in Lot II.

6. Each breeding cow required 1,090 pounds of straw for bedding, and produced 9,785 pounds of manure during the winter period.

7. The value of the manure more than paid for the cost of the labor and bedding required during the winter months.

8. The cost of wintering cows, during an average winter period of 154.6 days (5 months) under the conditions of this investigation, was more than twice the cost of pasturing them for 210.3 days (7 months).

---

[15] Based on actual cost of steers, The Annual Report, Pennsylvania State College, 1914-15.

[16] Bulletin 133, Pennsylvania Experiment Station.

9. The total cost of maintaining a beef cow in the Shorthorn lot was $33.54, and in the Aberdeen-Angus lot, $34.11.

10. The breeding results of the experiment indicated that the most important single factor for success in profitable beef production was regularity of the cows in producing calves. Selection of breeding cows for the herd should be made with as much emphasis on prolificacy as on form, quality, milking qualities, etc.

11. Under conditions of the experiment, when breeding 70 per cent. calves, they cost $49.73 and have a market value of $50.40.

12. The period of gestation was 285 days in both lots. In Lot I the calves averaged 75.27 pounds at birth and 671.6 pounds at twelve months of age, while the heifers at 24 months of age averaged 991.6 pounds. In Lot II (Angus) the calves averaged 68.36 pounds at birth and 588.8 pounds at 12 months, while the heifers at 24 months of age averaged 965 pounds.

13. The number of services per pregnancy in Lot I was 1.63 times and in Lot II, 1.43 times.

14. A pure-bred bull is essential. A herd of 50 to 60 cows is a unit for the best use of the sire, labor, equipment, and money invested (estimated).

15. Heifers at 30 months of age for breeding purposes cost less than their market value as beef.

16. Provided 80 per cent. calves are raised, steers under the conditions of the experiment would be produced at $6.10 per cwt. They were of better grade than steers purchased for feeding purposes by the college in the fall of 1913 that cost $7.45 per cwt. when placed in the feed lot.

17. This investigation shows that beef production can be made profitable under present Pennsylvania conditions where the breeding herd is maintained largely on roughage during the winter and on pasture during the summer, when cattle of insured beef type are used and when more than seventy per cent. calves are raised annually.

www.ingramcontent.com/pod-product-compliance
Lightning Source LLC
Chambersburg PA
CBHW062237220526
45471CB00009B/3521